574.5
R

The Question & Answer Book

ALL ABOUT PONDS

ALL ABOUT PONDS

By Jane Rockwell
Illustrated by Joseph Veno

Troll Associates

Library of Congress Cataloging in Publication Data

Rockwell, Jane.
  All about ponds.

  (Question and answer book)
  Summary: Answers questions about the stages in the
life of a pond and about the plants and animals that
may be found in and around ponds.
  1. Pond ecology—Miscellanea—Juvenile literature.
2. Ponds—Miscellanea—Juvenile literature. [1. Pond
ecology. 2. Ponds. 3. Ecology. 4. Questions and
answers] I. Veno, Joseph, ill. II. Title.
III. Series.
QH541.5.P63R63 1984   574.5'26322   83-4835
ISBN 0-89375-971-6
ISBN 0-89375-972-4 (pbk.)

Copyright © 1984 by Troll Associates, Mahwah, New Jersey

All rights reserved. No part of this book may be used
or reproduced in any manner whatsoever without written
permission from the publisher.

Printed in the United States of America
10   9   8   7   6   5   4   3   2   1

How does a pond begin?

There are many ways a pond may begin. This one started as a puddle one spring day. Melting snows and soft spring rains flowed down a slope and met at the bottom. More rains came. The puddle became a pool. An underground spring bubbled up and began to feed the pool. Soon, the pool became a pond.

How long will this pond live?

No one knows. It may come and go like the spring rains. Or it may last a very long time. If it lasts, it will go through several stages. It will slowly change from a young pond to a mature pond. Later, it will become an aging pond. And near the end of its life, it will become an old pond.

These changes will happen so slowly that if you lived near the pond and saw it every day, you might not even notice. But if you moved away and only came back to visit, you would notice how different the pond looks over the years.

What does a young pond look like?

Come visit this young pond. Take a good look. It is springtime. How quiet and peaceful everything seems. Suddenly, the splash of a diving frog ripples the clear, shining water.

This young pond is not very deep. Like most ponds, it is shallow enough so plants that take root in the muddy bottom can grow up to the surface. Other plants grow along the shoreline.

What is an ecosystem?

Beneath the still water of the pond, a busy, bustling world is awakening. There may be as many as a million plants and animals in this young pond! And each one is an important part of the pond *ecosystem*. An ecosystem is a community of living things. The plants and animals that live in the pond make up the pond ecosystem. They depend on one another to survive.

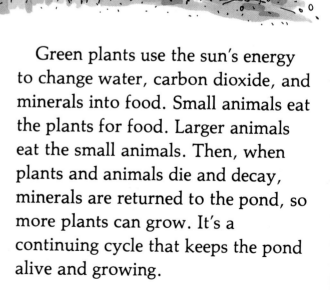

Green plants use the sun's energy to change water, carbon dioxide, and minerals into food. Small animals eat the plants for food. Larger animals eat the small animals. Then, when plants and animals die and decay, minerals are returned to the pond, so more plants can grow. It's a continuing cycle that keeps the pond alive and growing.

When the pond was first beginning to form, the bottom was clean. There were no plants or animals, because there was nothing for them to eat. But rainstorms washed soil and minerals into the pond, and before long, it was filled with tiny plants and animals.

What part does algae play in the ecosystem?

Look down into the water. You may see long strings of algae in this young pond. These simple, one-celled green plants are found in all ponds. Algae are the most important food producers for a pond community. As they make food, they also release oxygen into the water. This oxygen makes it possible for animals to survive in the water, since animals need oxygen in order to live.

Algae are producers, but some simple plants—like the bladderwort—are consumers. You may see some of these delicate-stemmed plants in the young pond. The bladderwort is an animal-eater. It traps tiny water animals and digests them.

What are the "invisible" animals in a pond?

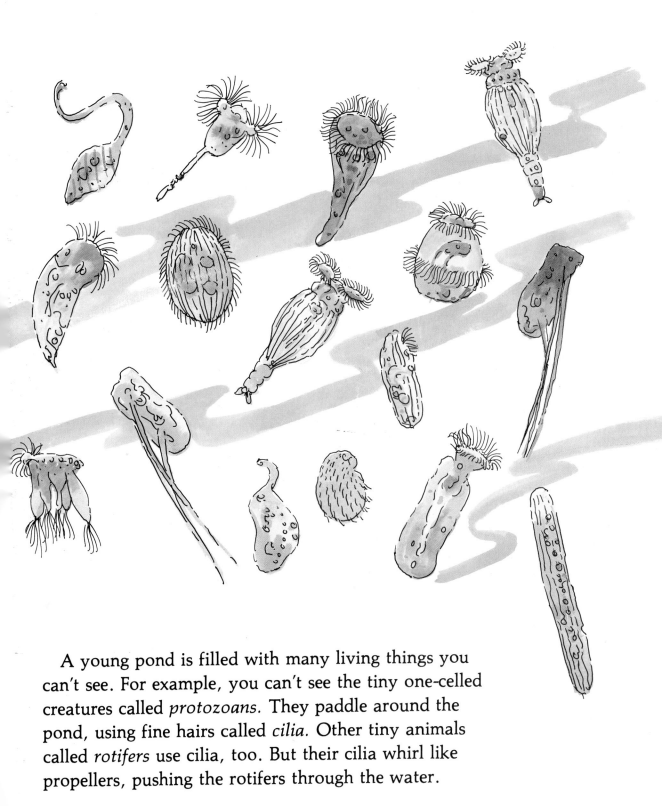

A young pond is filled with many living things you can't see. For example, you can't see the tiny one-celled creatures called *protozoans.* They paddle around the pond, using fine hairs called *cilia.* Other tiny animals called *rotifers* use cilia, too. But their cilia whirl like propellers, pushing the rotifers through the water.

What else lives here?

Wind and rain and melting ice and snow have brought many plants and animals to the young pond. Insects have come from bodies of water nearby. Birds and ducks have brought seeds and insect eggs on their webbed feet.

If you look closely, you might see a jelly-like mass in the shallow water. It is made up of frog's eggs. In the springtime, female frogs lay their eggs, and male frogs fertilize them. A frog may lay as many as 10,000 eggs in the pond. In less than two weeks, the eggs hatch into tiny, fish-like tadpoles. Some are so small that they can hardly be seen.

By summertime, if they haven't been eaten by fish or turtles, the tadpoles will have changed into frogs. Bullfrog tadpoles take about two years to develop into frogs. Then their strange call—*jug-o-rum, jug-o-rum*—will ring out loud and clear as night settles over the pond.

But now it is still spring, and you will not see the pond this summer. You will not see the pond slowly change. In later years, if you do return, this pond will not look the same.

What does a mature pond look like?

Come visit the pond, and see how it has changed. It is no longer young—now it is a mature pond. Perhaps it is summer, many years since your last visit.

There, at the edge of the shallow water, stands a great blue heron—one of the largest birds at the pond. It is more than one-and-a-half yards (1.4 meters) tall, from the tip of its long bill to its webbed feet. This gray-blue bird stands at the edge of the water, waiting for its dinner to swim by. When the heron sees its prey, it stretches out its long neck and stabs with its bill.

You can tell that this is a mature pond by looking at the bottom. The bottom of the pond looks like an underwater jungle. Plants also line the shore and have begun to creep into the pond itself. Bits of earth and rock called *sediment* have begun to build up along the shore and on the floor of the pond.

Has the ecosystem of the pond changed?

Yes. There are more kinds of plants and animals in this mature pond than when the pond was young. The ecosystem has changed. The population of the pond community has changed.

What animals live here?

This pond community is very busy. Noisy, quacking mallard ducks chase each other across the pond. A mother mallard swims calmly past the water lilies, followed by a line of fuzzy ducklings.

Mallards love to make a meal of the plants that grow in the pond. But they have a funny way of eating them. When a mallard eats plants in a pond, it seems to stand on its head. Only its tail feathers can be seen above water.

The pond is full of life on this warm summer day. There is a turtle, resting lazily in the sun. The pond has many water plants, insects, and snails for the turtle to eat.

On the other side of the pond, a kingfisher bird hovers like a helicopter over the water, looking for a meal. This bird, with its pretty blue-gray and white feathers, rarely goes hungry, for there are many fish in the mature pond. There are perch and sunfish and sometimes large-mouth bass—which may be a yard (1 meter) in length.

Curling in and out among the fish and water plants are water snakes. They are about the same length as the largemouth bass, and they have brightly colored bands on their bodies. Like other reptiles, water snakes are afraid of people and try to get away from them. But if they feel trapped, they will bite! The bite will hurt, but most water snakes aren't poisonous. They eat fish, frogs, and toads. Ribbon snakes live in ponds, too. They swim with their noses above the water as they curve along.

On warm summer evenings, the pond mosquitoes come out. The female deposits her eggs on the surface of the water. The eggs, which look like little black dots, will hatch in just a few hours. Out comes a wriggler— the first stage of the mosquito. Immediately, the wriggler starts to eat the green algae in the pond. In a few days, it will become an adult mosquito and will buzz away.

How do other ponds begin?

Nature started this pond several years ago, feeding it with flowing streams and an underground spring. But people can make ponds, too. Sometimes such ponds are used to water livestock or to irrigate land.

Other ponds may be built by beavers. These hard-working animals build themselves a home by cutting down trees with their strong teeth. Then they dam the waters of a stream with the trees, and strengthen the dam with rocks and mud.

In the middle of this pond, a beaver family builds its lodge. A beaver lodge is a home of sticks. Inside, there is a platform, or floor, that is above the level of the water in the pond. The entrance to the lodge is an underwater passageway.

What are the pond's night sounds?

A whole orchestra of sound surrounds this mature pond on a summer night. Mosquitoes and other flying insects buzz. A tree frog *peep-peeps*, and a fish makes a quiet splash as it breaks the surface of the water. A raccoon wades into the water, looking for fish and frogs.

Summer is a busy time in the world of a pond. But now you must leave. And while you are gone, this mature pond will continue to grow—and change.

What does an aging pond look like?

Come visit this aging pond. Perhaps it is autumn, several years since your last visit. There is a slight chill in the air on this clear, golden morning. A leaf turns red and yellow and flutters in the breeze. It drops to the surface of the pond and floats across, like a tiny boat.

The pond is growing older. More and more plants line the shore and creep out into the water. More sediment builds up on the bottom. More pond plants and animals die and decay. Their remains settle on the bottom of the pond, slowly filling it.

What will happen to the pond?

In time, if it is not cleared, the pond will become smaller and shallower, and the entire surface will be covered with plants. Then it may turn into a marsh or a swamp. If the aging pond has poor drainage, a peat bog may form. Layers of peat—decayed plant matter—may gradually pile up in the shallow water. In time, trees and shrubs may begin to grow in the rich peat.

This pond is aging, but it is not yet old. There is still life in its busy world. But winter is coming, and the animals of the pond know it.

Some of the birds are leaving the pond to begin their journey south. The mosquitoes are gone. The tadpoles have turned into frogs and hopped out of the water to live on the land.

The beavers are busy filling their lodge with food for the coming winter. They also store food underwater nearby. They cut tender twigs and stick them into the mud at the bottom of the pond. Later, when the surface of the pond freezes, they can leave their lodge through the underwater passageway, and return with a tasty meal.

Where do ponds form?

Some ponds, like this one, are in a forest clearing or in the middle of a rich meadow. But ponds can also form in many other places. If sand dunes grow tall enough at the ocean shore, they can trap enough water to form saltwater ponds. Other ponds may be formed by meandering streams. A meandering stream curves back and forth across the land. If one of the curved sections is cut off from the rest of the stream, the cut-off part may become a pond.

Each pond has a life of its own. And each pond must change as it grows. A young pond matures. A mature pond ages. And an aging pond grows old.

What does an old pond look like?

Come visit this pond when it is old. The pond seems deserted on this winter morning. The air is crisp and cold, and most of the birds are far away, in warmer lands. The beavers have abandoned their lodge and moved to a younger pond, where there is more room. But not all of the pond's creatures have gone. Many are *hibernating*, sleeping through the cold winter months.

Where do pond animals hibernate?

Snakes crawl into rotted logs or under stones for a long winter's rest. Frogs and turtles dig holes in the mud, where they settle down for the winter. Frogs have blood vessels under their loose, moist skin that can absorb moisture and oxygen while they are hibernating. Turtles can take in air and water while they hibernate. They have openings in their skin that act like a fish's gills.

What are the signs of an old pond?

This pond has grown old. The water lilies and other floating plants have made bridges from one shore to the other. There is much less water now, because the bottom has grown thick with sediment. Perhaps, when summer comes again, the sun will dry up the pond completely. Some of the frogs and turtles have already left, seeking clearer, more open water.

The pond is filling up with silt. The shoreline grows out into the pond more and more, and the plants form a thick cover that chokes off most of the sun's rays. Little by little, the pond is dying.

What will take the place of the pond?

Tiny one-celled plants and animals can stay alive as long as there is water and sunlight. Crayfish and leeches and snails can survive by burrowing into damp soil or by finding moisture under densely growing plants.

Still, it is only a question of time until this old pond and its creatures are no more. In their place may be a lush, green meadow. And in more time, the meadow will change, too. It may become a thicket of shrubs. Then trees may take root and eventually become a forest.

Can you recall the changes?

Now, looking at this crowded, murky pond, can you remember how it looked when it was young? Do you recall seeing the bottom of the pond through crystal clean water? That was long ago. The people who live nearby hardly noticed the change. It happened slowly—over many years. But you noticed the changes.

With each visit to the pond, it seemed as if you were looking at a different world. And in a way, you were.

At each stage in its life, a pond really is a new and different world. And every pond—whether it is young, mature, aging, or old—is unique and fascinating in its own special way.